Managing Editor
Karen Goldfluss, M.S. Ed.

Editor-in-Chief
Sharon Coan, M.S. Ed.

Cover Artist
Barb Lorseyedi

Art Coordinator
Kevin Barnes

Art Director
CJae Froshay

Imaging
James Edward Grace

Product Manager
Phil Garcia

Geometry

GRADES 3 & 4

W9-BEZ-379

Authors

Teacher Created Materials Staff

Publishers
Rachelle Cracchiolo, M.S. Ed.
Mary Dupuy Smith, M.S. Ed.

Teacher Created Materials, Inc.
6421 Industry Way
Westminster, CA 92683
www.teachercreated.com

ISBN-0-7439-3327-3

©2002 Teacher Created Materials, Inc.
Reprinted, 2004

Made in U.S.A.

Table of Contents

Introduction

The old adage "practice makes perfect" can really hold true for your child and his or her education. The more practice and exposure your child has with concepts being taught in school, the more success he or she is likely to find. For many parents, knowing how to help your children can be frustrating because the resources may not be readily available. As a parent it is also difficult to know where to focus your efforts so that the extra practice your child receives at home supports what he or she is learning in school.

This book has been designed to help parents and teachers reinforce basic skills with their children. *Practice Makes Perfect* reviews basic math skills for children in grades 3 and 4. The math focus is geometry. While it would be impossible to include all concepts taught in grades 3 and 4 in this book, the following basic objectives are reinforced through practice exercises. These objectives support math standards established on a district, state, or national level. (Refer to the Table of Contents for the specific objectives of each practice page.)

- identifying points, lines, segments, and rays
- identifying parallel, intersecting, and perpendicular lines
- describing and classifying angles
- identifying congruent and similar figures
- drawing lines of symmetry
- identifying flips, turns, and slides
- describing triangles
- measuring angles in a triangle
- naming plane and solid geometric figures
- identifying faces, edges, and vertices
- finding perimeter and area of polygons and triangles
- finding the radius and diameter of a circle

There are 36 practice pages organized sequentially, so children can build their knowledge from more basic skills to higher-level math skills. (**Note:** Have children show all work where computation is necessary to solve a problem. For multiple choice responses on practice pages, children can fill in the letter choice or circle the answer.) Following the practice pages are six practice tests. These provide children with multiple-choice test items to help prepare them for standardized tests administered in schools. As your child completes each test, he or she should fill in the correct bubbles on the answer sheet (page 46). To correct the test pages and the practice pages in this book, use the answer key provided on pages 47 and 48.

How to Make the Most of This Book

Here are some useful ideas for optimizing the practice pages in this book:

- Set aside a specific place in your home to work on the practice pages. Keep it neat and tidy with materials on hand.
- Set up a certain time of day to work on the practice pages. This will establish consistency. An alternative is to look for times in your day or week that are less hectic and conducive to practicing skills.
- Keep all practice sessions with your child positive and constructive. If the mood becomes tense, or you and your child are frustrated, set the book aside and look for another time to practice with your child.
- Help with instructions if necessary. If your child is having difficulty understanding what to do or how to get started, work through the first problem through with him or her.
- Review the work your child has done. This serves as reinforcement and provides further practice.
- Allow your child to use whatever writing instruments he or she prefers. For example, colored pencils can add variety and pleasure to drill work.
- Pay attention to the areas in which your child has the most difficulty. Provide extra guidance and exercises in those areas. Allowing children to use drawings and manipulatives, such as coins, tiles, game markers, or flash cards, can help them grasp difficult concepts more easily.
- Look for ways to make real-life applications to the skills being reinforced.

Practice 1

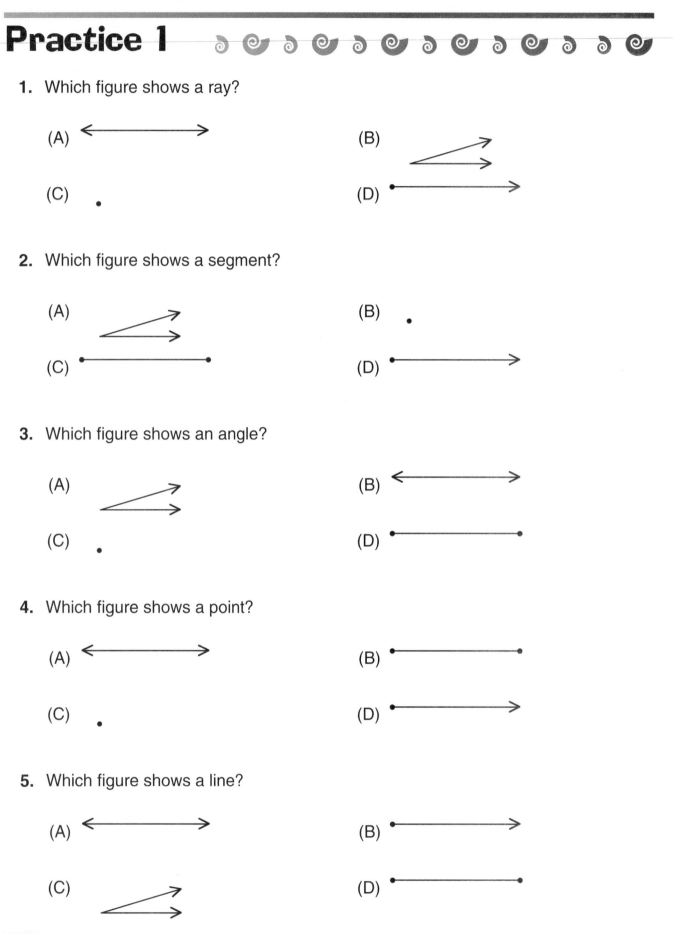

1. Which figure shows a ray?

   (A)   (B)

   (C)   (D)

2. Which figure shows a segment?

   (A)   (B)

   (C)   (D)

3. Which figure shows an angle?

   (A)   (B)

   (C)   (D)

4. Which figure shows a point?

   (A)   (B)

   (C)   (D)

5. Which figure shows a line?

   (A)   (B)

   (C)   (D)

Practice 2

1. Which word best describes the figure?

    •

    (A) line   (B) point   (C) ray   (D) angle

2. Which word best describes the figure?

    ⟶

    (A) ray   (B) point   (C) angle   (D) segment

3. Which word best describes the figure?

    ⟷

    (A) segment   (B) ray   (C) line   (D) angle

4. Which word best describes the figure?

    •———•

    (A) segment   (B) ray   (C) line   (D) point

5. Which word best describes the figure?

    (A) point   (B) segment   (C) line   (D) angle

Practice 3

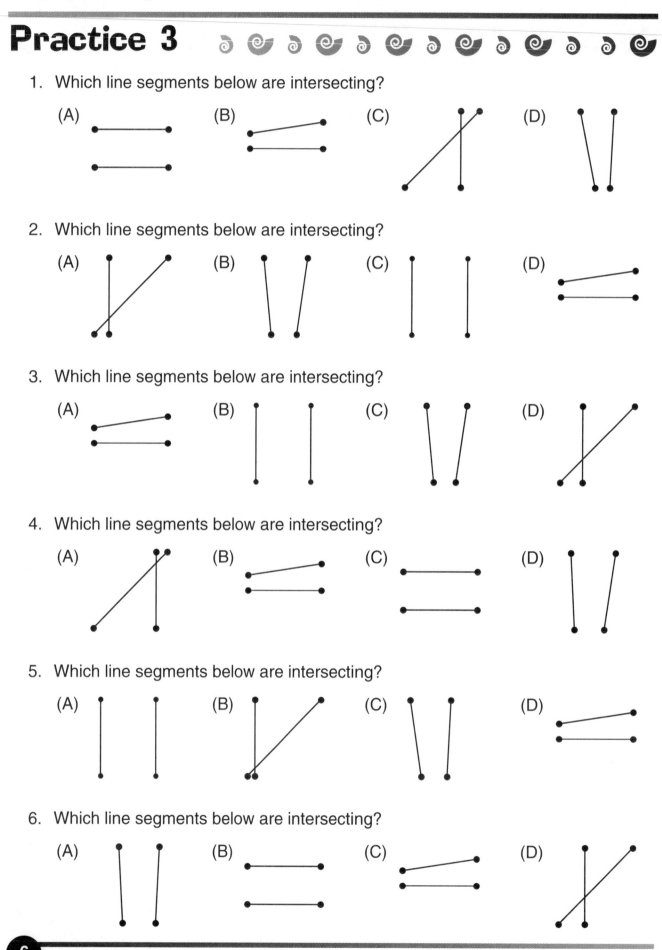

1. Which line segments below are intersecting?

    (A)   (B)   (C)   (D)

2. Which line segments below are intersecting?

    (A)   (B)   (C)   (D)

3. Which line segments below are intersecting?

    (A)   (B)   (C)   (D)

4. Which line segments below are intersecting?

    (A)   (B)   (C)   (D)

5. Which line segments below are intersecting?

    (A)   (B)   (C)   (D)

6. Which line segments below are intersecting?

    (A)   (B)   (C)   (D)

#3327 Practice Makes Perfect: Geometry

Practice 4

1. Which line segments below appear to be parallel?

   (A)   (B)   (C)   (D)

2. Which line segments below appear to be parallel?

   (A)   (B)   (C)   (D)

3. Which line segments below appear to be parallel?

   (A)   (B)   (C)   (D)

4. Which line segments below appear to be parallel?

   (A)   (B)   (C)   (D)

5. Which line segments below appear to be parallel?

   (A)   (B)   (C)   (D)

6. Which line segments below appear to be parallel?

   (A)   (B)   (C)   (D)

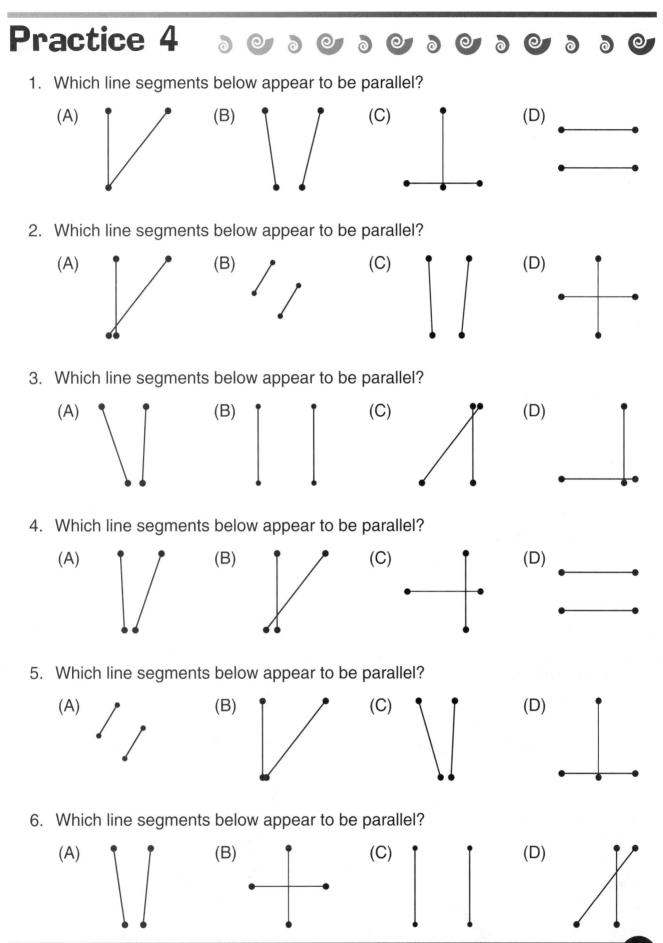

Practice 5

1. Describe the angle *a* in the figure.

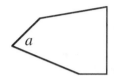

(A) an acute angle

(B) an obtuse angle

(C) a right angle

2. Describe the angle *a* in the figure.

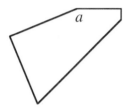

(A) an acute angle

(B) an obtuse angle

(C) a right angle

3. Describe the angle *a* in the figure.

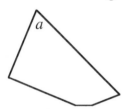

(A) an obtuse angle

(B) an acute angle   (C) a right angle

4. Describe the angle *a* in the figure.

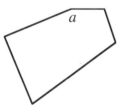

(A) an obtuse angle

(B) a right angle   (C) an acute angle

5. Describe the angle *a* in the figure.

(A) an acute angle

(B) an obtuse angle

(C) a right angle

6. Describe the angle *a* in the figure.

(A) an obtuse angle

(B) a right angle   (C) an acute angle

7. Describe the angle *a* in the figure.

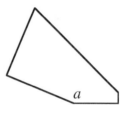

(A) an obtuse angle

(B) an acute angle   (C) a right angle

8. Describe the angle *a* in the figure.

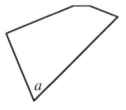

(A) an acute angle

(B) a right angle   (C) an obtuse angle

Practice 6

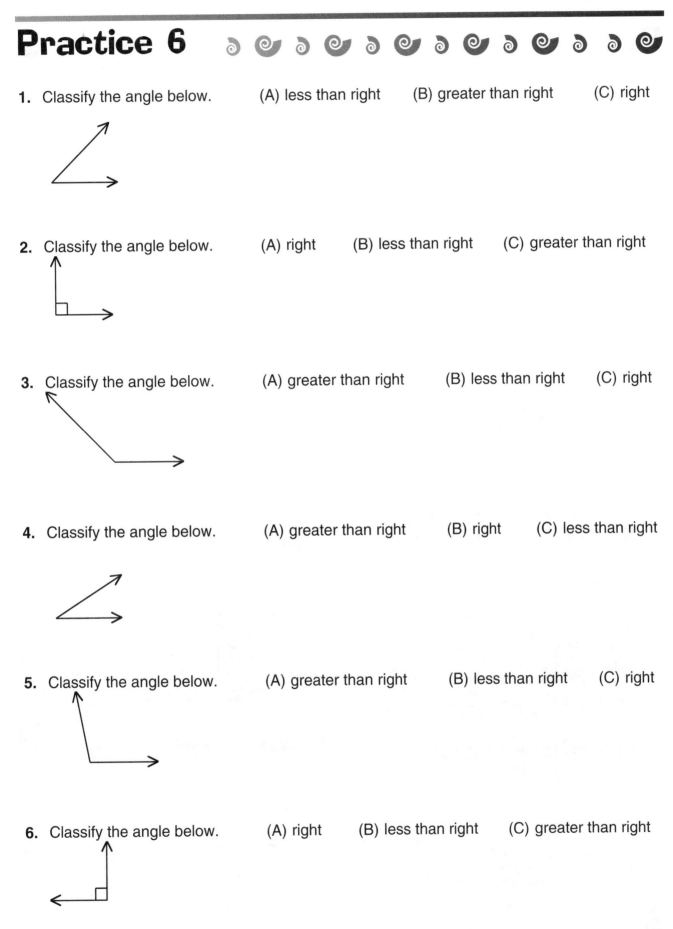

1. Classify the angle below.   (A) less than right   (B) greater than right   (C) right

2. Classify the angle below.   (A) right   (B) less than right   (C) greater than right

3. Classify the angle below.   (A) greater than right   (B) less than right   (C) right

4. Classify the angle below.   (A) greater than right   (B) right   (C) less than right

5. Classify the angle below.   (A) greater than right   (B) less than right   (C) right

6. Classify the angle below.   (A) right   (B) less than right   (C) greater than right

Practice 7

Write *Yes* or *No* next to the question.

1. Are the two figures the same shape and size?

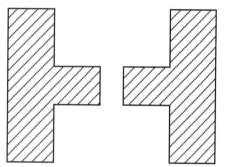

2. Are the two figures the same shape and size?

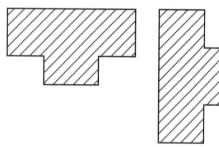

3. Are the two figures the same shape and size?

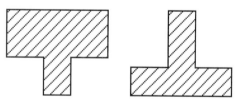

4. Are the two figures the same shape and size?

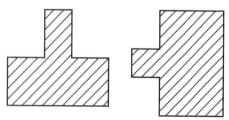

5. Are the two figures the same shape and size?

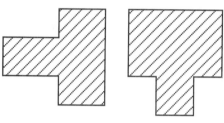

6. Are the two figures the same shape and size?

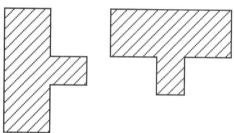

7. Are the two figures the same shape and size?

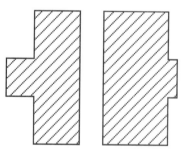

8. Are the two figures the same shape and size?

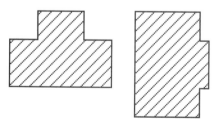

Practice 8

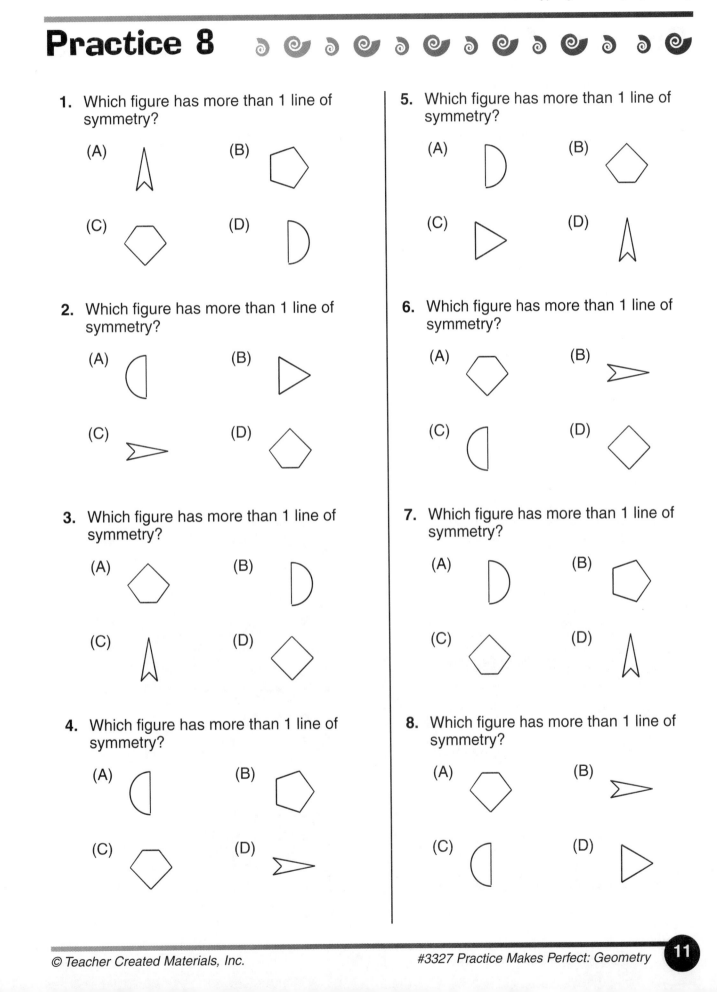

1. Which figure has more than 1 line of symmetry?

   (A)   (B)

   (C)   (D)

2. Which figure has more than 1 line of symmetry?

   (A)   (B)

   (C)   (D)

3. Which figure has more than 1 line of symmetry?

   (A)   (B)

   (C)   (D)

4. Which figure has more than 1 line of symmetry?

   (A)   (B)

   (C)   (D)

5. Which figure has more than 1 line of symmetry?

   (A)   (B)

   (C)   (D)

6. Which figure has more than 1 line of symmetry?

   (A)   (B)

   (C)   (D)

7. Which figure has more than 1 line of symmetry?

   (A)   (B)

   (C)   (D)

8. Which figure has more than 1 line of symmetry?

   (A)   (B)

   (C)   (D)

Practice 9

1. Which shape has a line of symmetry?

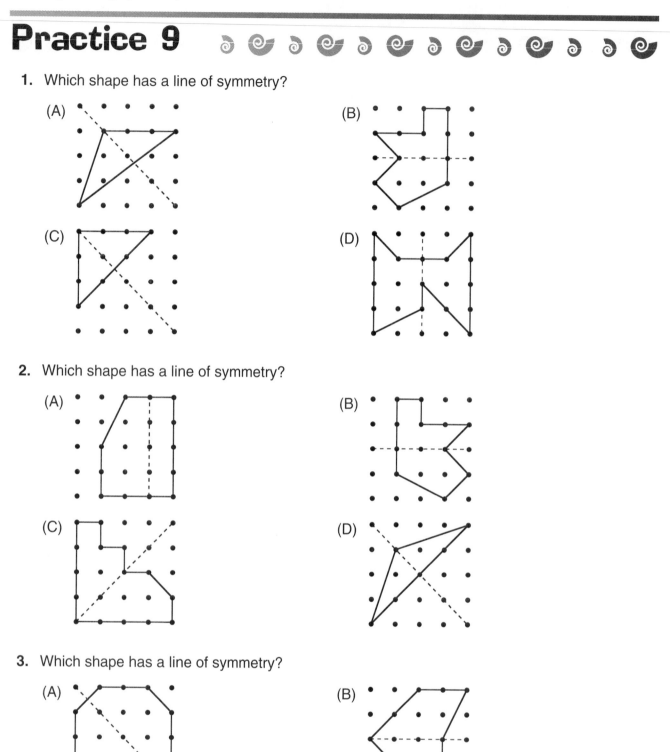

(A)

(B)

(C)

(D)

2. Which shape has a line of symmetry?

(A)

(B)

(C)

(D)

3. Which shape has a line of symmetry?

(A)

(B)

(C)

(D)

Practice 10

1. Which of the following shows a flip?

(A)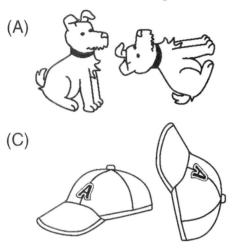

(B)

(C)

(D) none of these

2. Which of the following shows a slide?

(A)

(B)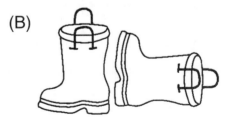

(C)

(D) none of these

3. Which of the following shows a turn?

(A)

(B)

(C)

(D) none of these

Practice 11

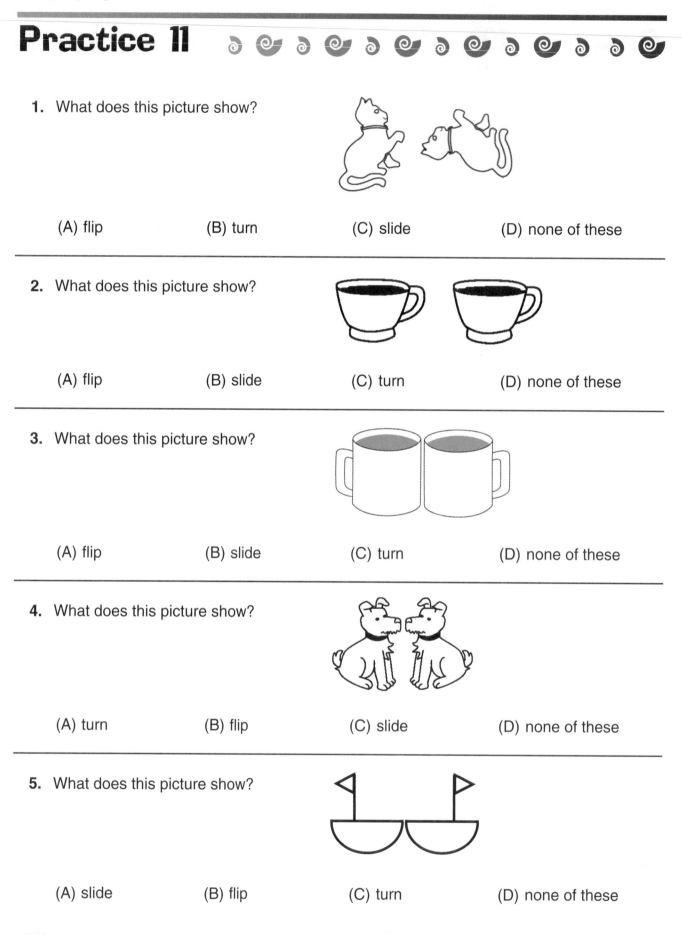

1. What does this picture show?

   (A) flip   (B) turn   (C) slide   (D) none of these

2. What does this picture show?

   (A) flip   (B) slide   (C) turn   (D) none of these

3. What does this picture show?

   (A) flip   (B) slide   (C) turn   (D) none of these

4. What does this picture show?

   (A) turn   (B) flip   (C) slide   (D) none of these

5. What does this picture show?

   (A) slide   (B) flip   (C) turn   (D) none of these

Practice 12

Write whether the picture shows a *slide*, a *flip*, or a *turn*.

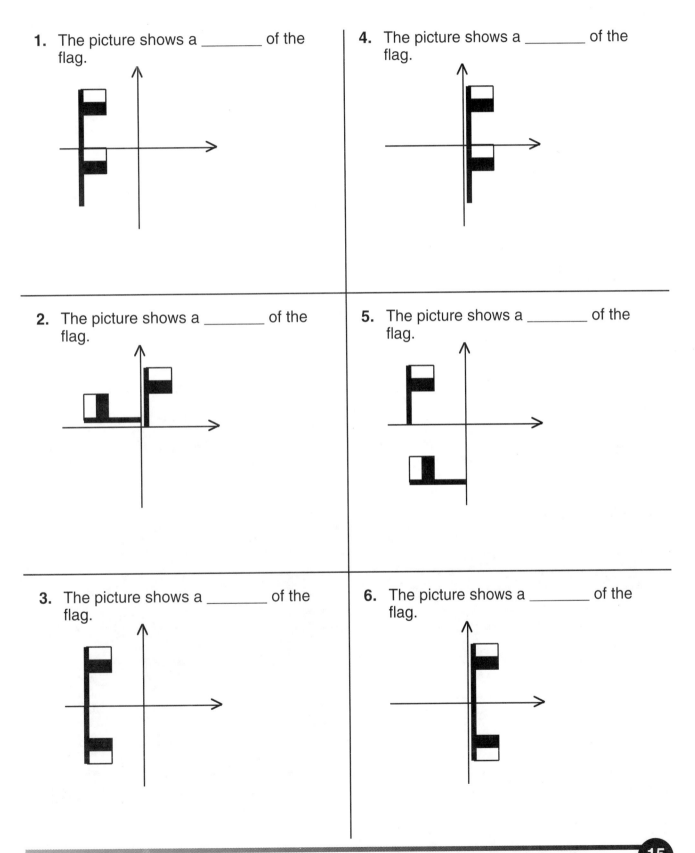

1. The picture shows a _____ of the flag.

4. The picture shows a _____ of the flag.

2. The picture shows a _____ of the flag.

5. The picture shows a _____ of the flag.

3. The picture shows a _____ of the flag.

6. The picture shows a _____ of the flag.

Practice 13

1. Look at this figure.

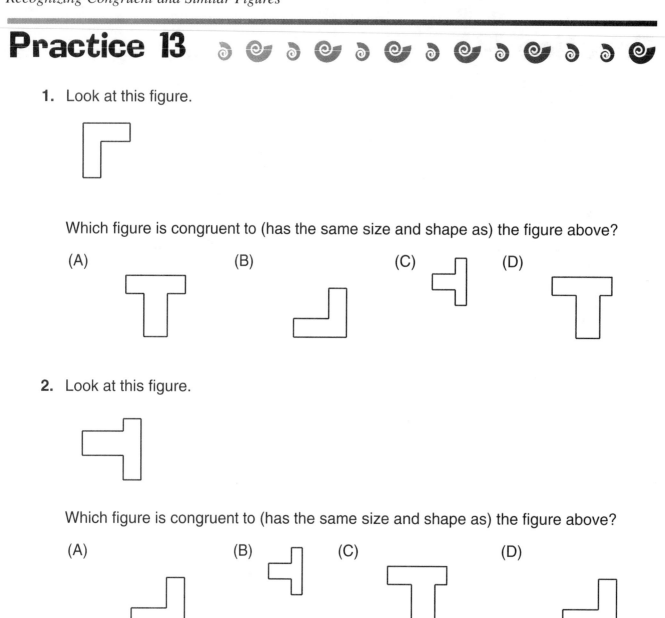

Which figure is congruent to (has the same size and shape as) the figure above?

(A)　　　　　(B)　　　　　(C)　　　　　(D)

2. Look at this figure.

Which figure is congruent to (has the same size and shape as) the figure above?

(A)　　　　　(B)　　　　　(C)　　　　　(D)

3. Look at this figure.

Which figure is congruent to (has the same size and shape as) the figure above?

(A)　　　　　(B)　　　　　(C)　　　　　(D)

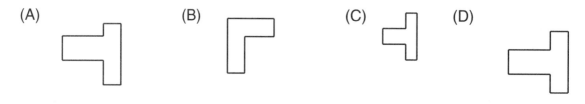

Practice 14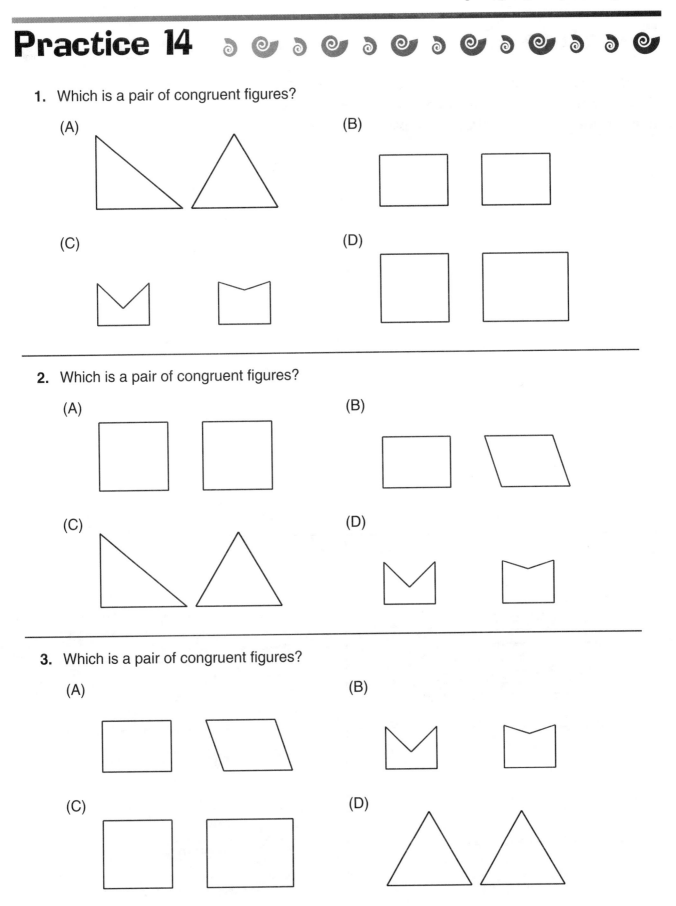

1. Which is a pair of congruent figures?

(A)

(B)

(C)

(D)

2. Which is a pair of congruent figures?

(A)

(B)

(C)

(D)

3. Which is a pair of congruent figures?

(A)

(B)

(C)

(D)

Practice 15

An **acute** angle is less than 90°. A **right** angle is exactly 90°. An **obtuse** angle is more than 90° and less than 180°. A **straight** angle is exactly 180°. Label each of the angles below as **acute**, **right**, **obtuse**, or **straight** angles.

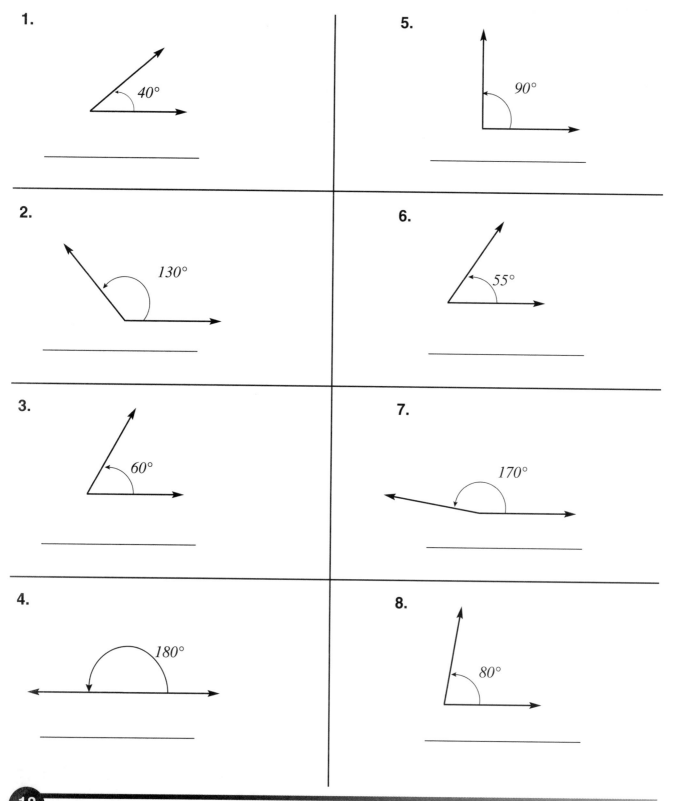

1.

40°

_____

2.

130°

_____

3.

60°

_____

4.

180°

_____

5.

90°

_____

6.

55°

_____

7.

170°

_____

8.

80°

_____

Practice 16 ⟡ ⟡ ⟡ ⟡ ⟡ ⟡ ⟡ ⟡ ⟡ ⟡ ⟡ ⟡ ⟡ ⟡

Read the information in the box to decide whether each triangle below is **right**, **equilateral, isosceles, scalene, isosceles right**, **acute**, or **obtuse**. (Some triangles may be more than one type.)

- A *right* triangle has one 90° angle.
- An *equilateral* triangle has three equal sides and three equal angles of 60° each.
- An *isosceles* triangle has two equal sides and two equal angles.
- A *scalene* triangle has no equal sides and no equal angles.
- An *isosceles right* triangle has one 90° angle and two 45° angles. The sides adjacent (next to) the right angle are equal.
- An *acute* triangle has all three angles less than 90°.
- An *obtuse* triangle has one angle greater than 90°.

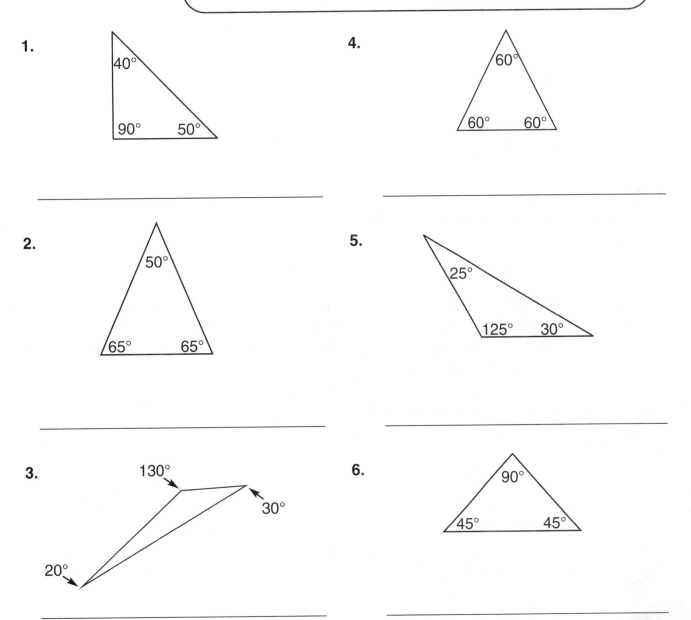

1.

40°
90° 50°

4.

60°
60° 60°

2.

50°
65° 65°

5.

25°
125° 30°

3.

130°
30°
20°

6.

90°
45° 45°

Practice 17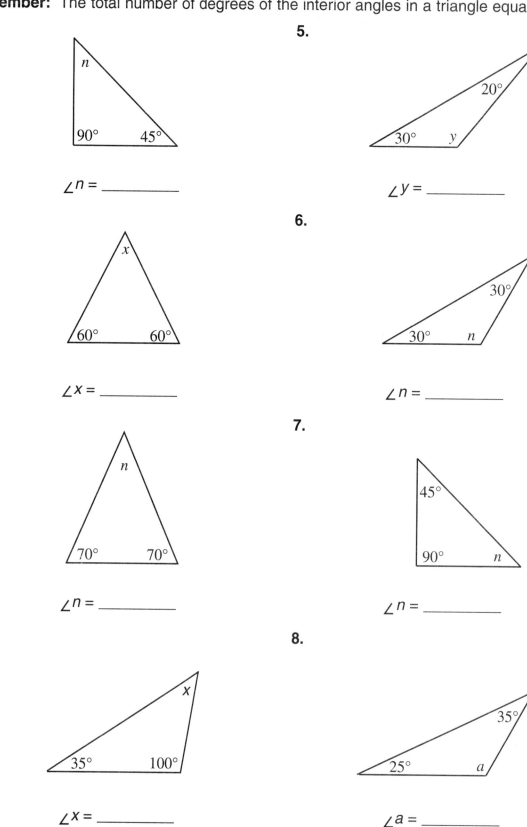

Find the number of degrees in each unmarked angle.

Remember: The total number of degrees of the interior angles in a triangle equals 180.

1.

n

$90°$ $45°$

$\angle n =$ _____

2.

x

$60°$ $60°$

$\angle x =$ _____

3.

n

$70°$ $70°$

$\angle n =$ _____

4.

x

$35°$ $100°$

$\angle x =$ _____

5.

$20°$

$30°$ y

$\angle y =$ _____

6.

$30°$

$30°$ n

$\angle n =$ _____

7.

$45°$

$90°$ n

$\angle n =$ _____

8.

$35°$

$25°$ a

$\angle a =$ _____

Practice 18

1. Which figure is a kite?

    (A) 　　　　　(B) 　　　　　(C) 　　　　　(D)

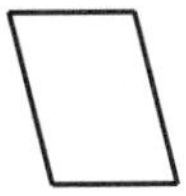

2. Which figure is a square?

    (A) 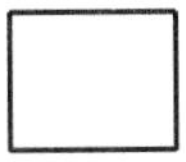　　　　　(B) 　　　　　(C)　　　　　(D)

3. Which figure is a rectangle?

    (A) 　　　　　(B) 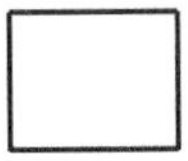　　　　　(C) 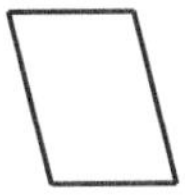　　　　　(D)

4. Which figure is a parallelogram?

    (A) 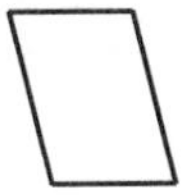　　　　　(B) 　　　　　(C) 　　　　　(D)

5. Which figure is a trapezoid?

    (A) 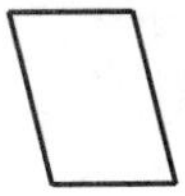　　　　　(B) 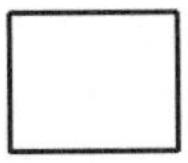　　　　　(C) 　　　　　(D)

6. Which figure is a triangle?

    (A) 　　　　　(B) 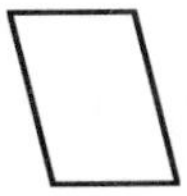　　　　　(C) 　　　　　(D)

Practice 19

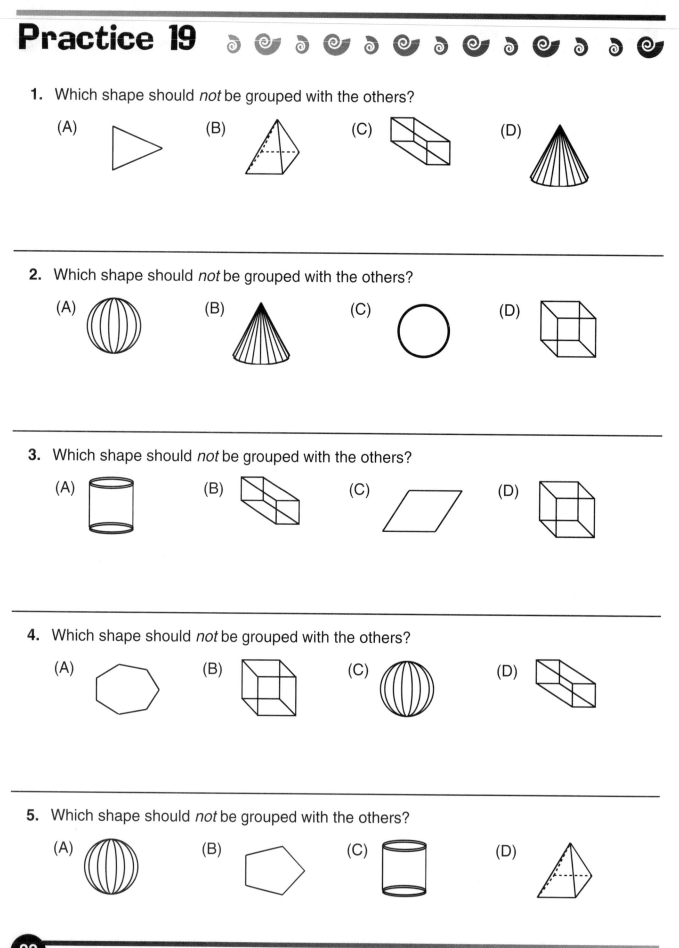

1. Which shape should *not* be grouped with the others?

   (A) (B) (C) (D)

2. Which shape should *not* be grouped with the others?

   (A) (B) (C) (D)

3. Which shape should *not* be grouped with the others?

   (A) (B) (C) (D)

4. Which shape should *not* be grouped with the others?

   (A) (B) (C) (D)

5. Which shape should *not* be grouped with the others?

   (A) (B) (C) (D)

Practice 20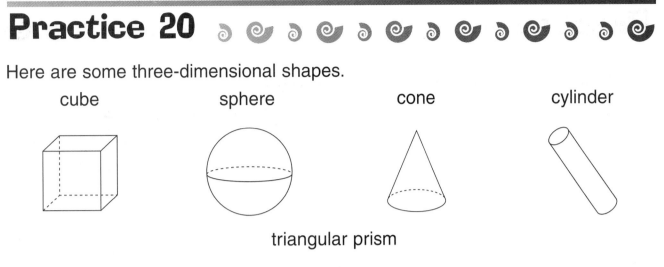

Here are some three-dimensional shapes.

cube   sphere   cone   cylinder

triangular prism

Write the name of the shape that matches the following objects.

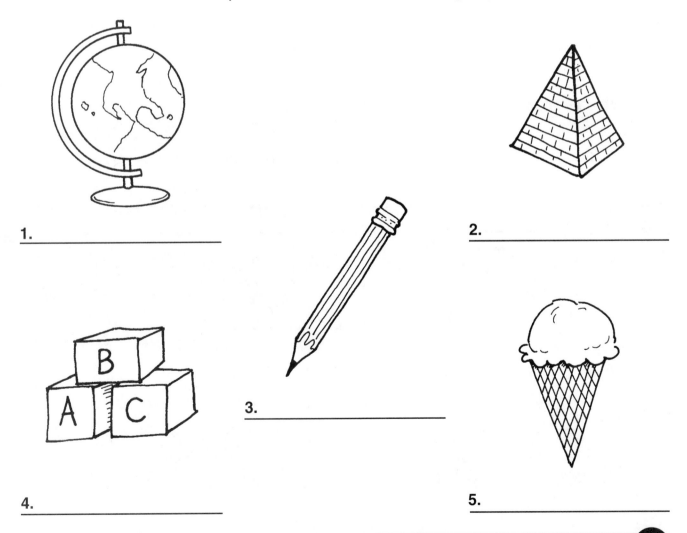

1. _____

2. _____

3. _____

4. _____

5. _____

Practice 21

1.

cube	rectangular prism	cylinder	sphere	cone	pyramid

Which solid does the die look like?

(A) cylinder　　(B) square　　(C) cube　　(D) sphere

2.

cube	rectangular prism	cylinder	sphere	cone	pyramid

Which solid does the stick of butter look like?

BUTTER

(A) cube　　(B) rectangle　　(C) rectangular prism　　(D) square

3.

cube	rectangular prism	cylinder	sphere	cone	pyramid

Which solid does the ball look like?

(A) cone　　(B) pyramid　　(C) cylinder　　(D) sphere

Practice 22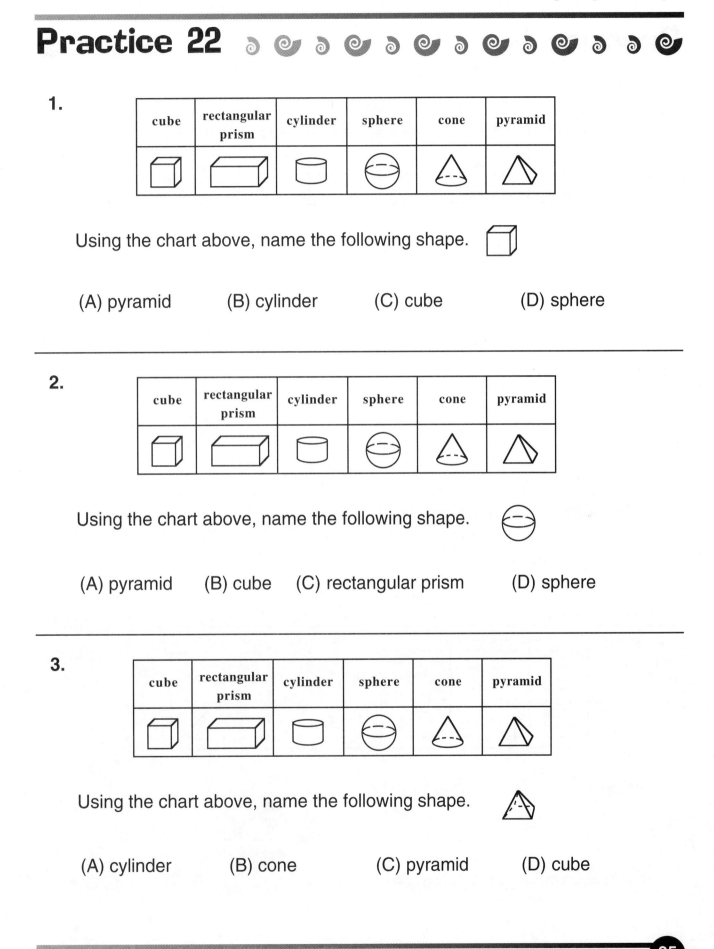

1.

cube	rectangular prism	cylinder	sphere	cone	pyramid
□	▱	⬭	◯	△	△

Using the chart above, name the following shape.

(A) pyramid   (B) cylinder   (C) cube   (D) sphere

2.

cube	rectangular prism	cylinder	sphere	cone	pyramid
□	▱	⬭	◯	△	△

Using the chart above, name the following shape.

(A) pyramid   (B) cube   (C) rectangular prism   (D) sphere

3.

cube	rectangular prism	cylinder	sphere	cone	pyramid
□	▱	⬭	◯	△	△

Using the chart above, name the following shape.

(A) cylinder   (B) cone   (C) pyramid   (D) cube

Practice 23

1.

cube	rectangular prism	cylinder	sphere	cone	pyramid

Which word best describes this figure?

_____

2.

cube	rectangular prism	cylinder	sphere	cone	pyramid

Which word best describes this figure?

_____

3.

cube	rectangular prism	cylinder	sphere	cone	pyramid

Which word best describes this figure?

_____

Practice 24

Count the number of faces, edges, and verticles for each geometric solid on this page. Use the following list to name the figure.

tetrahedron

hexahedron

octahedron

triangular prism

icosahedron

square pyramid

1.

name _____

faces _____

edges_____

vertices _____

4.

name _____

faces _____

edges _____

vertices _____

2.

name _____

faces _____

edges_____

vertices _____

5.

name _____

faces _____

edges _____

vertices _____

3.

name _____

faces _____

edges_____

vertices _____

6.

name _____

faces _____

edges _____

vertices _____

Practice 25 🐚 🐚 🐚 🐚 🐚 🐚 🐚 🐚 🐚 🐚 🐚 🐚 🐚

Use the list below to identify each of these figures.

hexahedron   **cylinder**   **dodecahedron**

octahedron   **cone**   **square pyramid**

triangular pyramid   **sphere**

1. _____

2. _____

3. _____

4. _____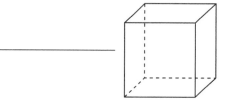

5. _____

6. _____

7. _____

8. _____

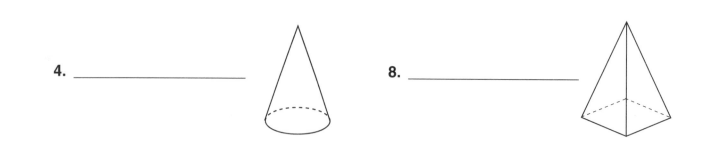

Practice 26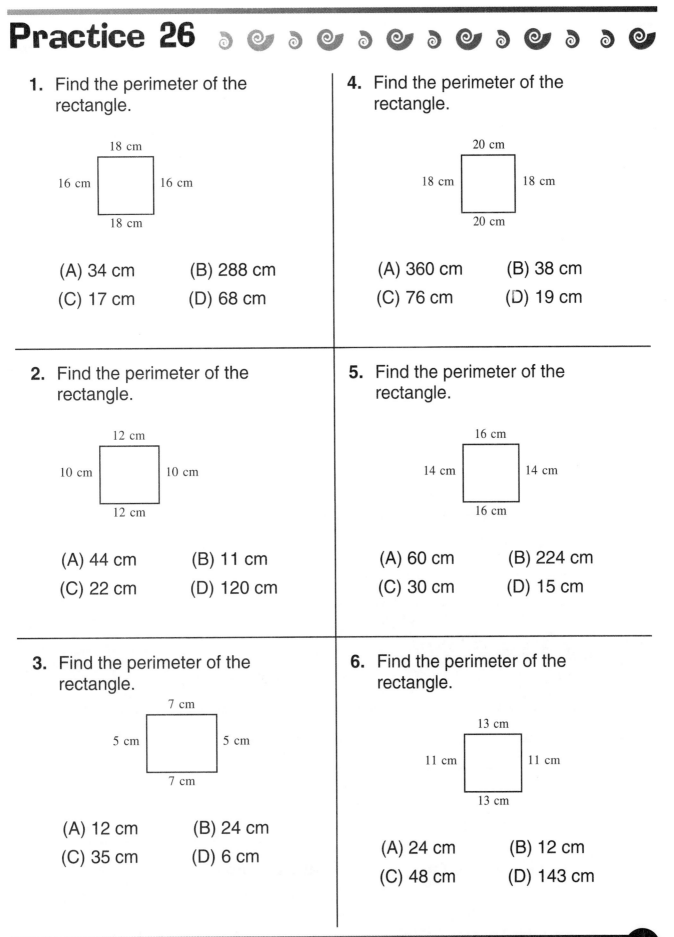

1. Find the perimeter of the rectangle.

18 cm
16 cm   16 cm
18 cm

(A) 34 cm   (B) 288 cm

(C) 17 cm   (D) 68 cm

2. Find the perimeter of the rectangle.

12 cm
10 cm   10 cm
12 cm

(A) 44 cm   (B) 11 cm

(C) 22 cm   (D) 120 cm

3. Find the perimeter of the rectangle.

7 cm
5 cm   5 cm
7 cm

(A) 12 cm   (B) 24 cm

(C) 35 cm   (D) 6 cm

4. Find the perimeter of the rectangle.

20 cm
18 cm   18 cm
20 cm

(A) 360 cm   (B) 38 cm

(C) 76 cm   (D) 19 cm

5. Find the perimeter of the rectangle.

16 cm
14 cm   14 cm
16 cm

(A) 60 cm   (B) 224 cm

(C) 30 cm   (D) 15 cm

6. Find the perimeter of the rectangle.

13 cm
11 cm   11 cm
13 cm

(A) 24 cm   (B) 12 cm

(C) 48 cm   (D) 143 cm

Practice 27

1. Which answer gives the perimeter of the figure?

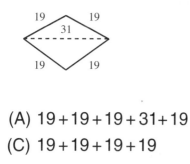

(A) $2 \times 19 + 2 \times 28$ (B) 19×28
(C) $19 + 19 + 19 + 19$ (D) $19 + 19 + 19 + 28 + 19$

2. Which answer gives the perimeter of the figure?

(A) $19 + 19 + 19 + 31 + 19$ (B) 19×31
(C) $19 + 19 + 19 + 19$ (D) $2 \times 19 + 2 \times 31$

3. Which answer gives the perimeter of the figure?

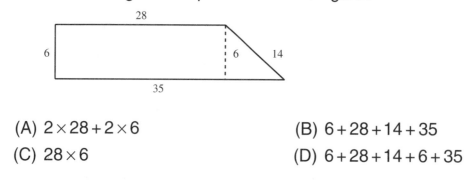

(A) $2 \times 28 + 2 \times 6$ (B) $6 + 28 + 14 + 35$
(C) 28×6 (D) $6 + 28 + 14 + 6 + 35$

4. Which answer gives the perimeter of the figure?

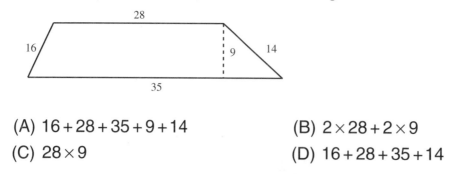

(A) $16 + 28 + 35 + 9 + 14$ (B) $2 \times 28 + 2 \times 9$
(C) 28×9 (D) $16 + 28 + 35 + 14$

Practice 28

Find the perimeter of each of these figures.

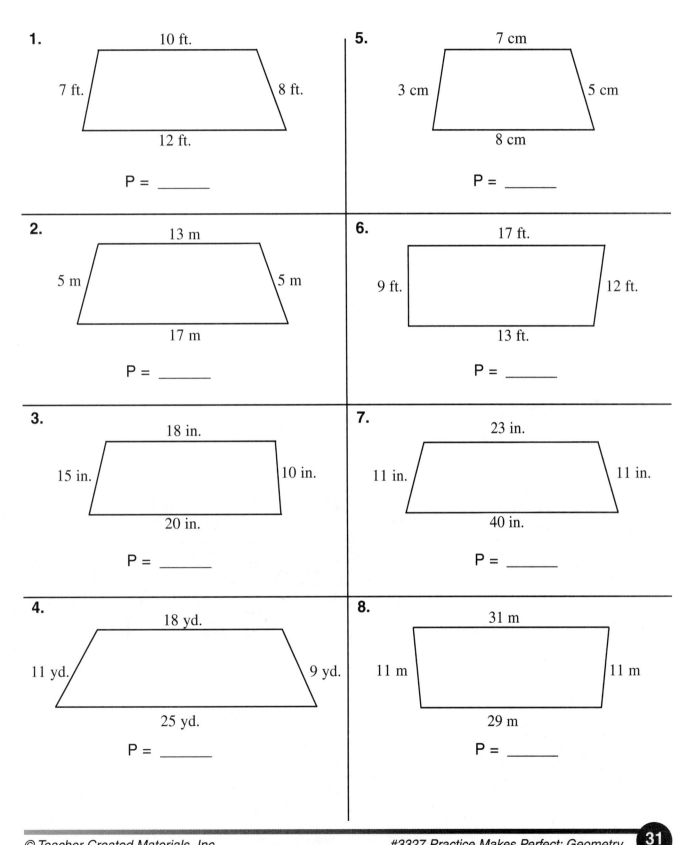

1.

10 ft.

7 ft.   8 ft.

12 ft.

P = _____

2.

13 m

5 m   5 m

17 m

P = _____

3.

18 in.

15 in.   10 in.

20 in.

P = _____

4.

18 yd.

11 yd.   9 yd.

25 yd.

P = _____

5.

7 cm

3 cm   5 cm

8 cm

P = _____

6.

17 ft.

9 ft.   12 ft.

13 ft.

P = _____

7.

23 in.

11 in.   11 in.

40 in.

P = _____

8.

31 m

11 m   11 m

29 m

P = _____

Practice 29

A regular polygon has equal sides. Find the perimeter of each of these figures.

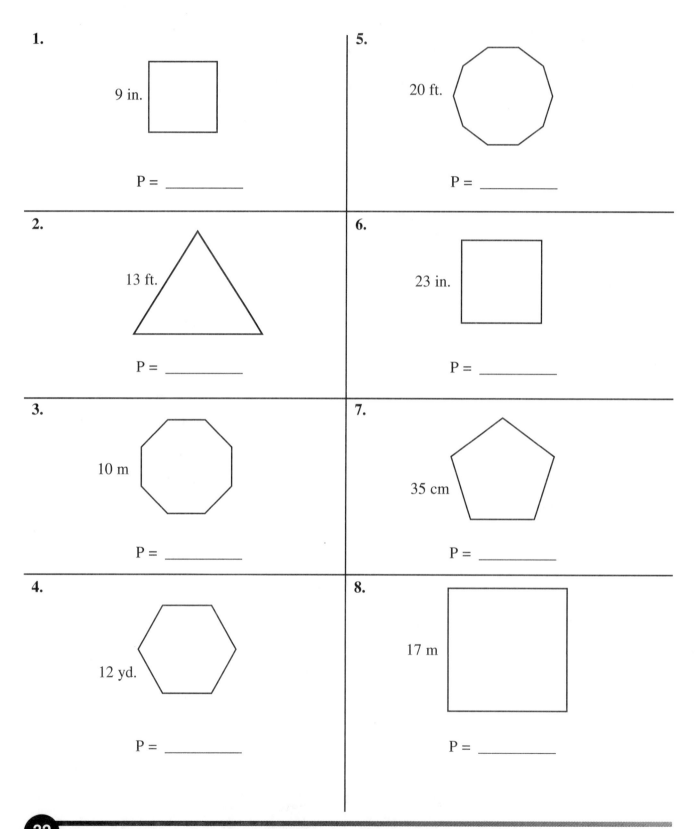

1.

9 in.

P = _____

2.

13 ft.

P = _____

3.

10 m

P = _____

4.

12 yd.

P = _____

5.

20 ft.

P = _____

6.

23 in.

P = _____

7.

35 cm

P = _____

8.

17 m

P = _____

Practice 30

1. What is the area of the figure drawn on the grid?

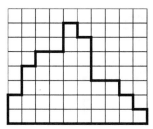

   (A) 38 square units

   (B) 39 square units

   (C) 36 square units

   (D) 41 square units

2. What is the area of the figure drawn on the grid?

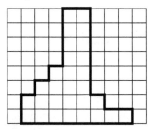

   (A) 29 square units

   (B) 28 square units

   (C) 27 square units

   (D) 32 square units

3. What is the area of the figure drawn on the grid?

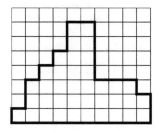

   (A) 38 square units

   (B) 40 square units

   (C) 35 square units

   (D) 37 square units

4. What is the area of the figure drawn on the grid?

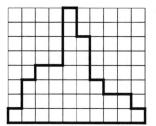

   (A) 34 square units

   (B) 31 square units

   (C) 33 square units

   (D) 36 square units

5. What is the area of the figure drawn on the grid?

   (A) 43 square units

   (B) 40 square units

   (C) 45 square units

   (D) 42 square units

6. What is the area of the figure drawn on the grid?

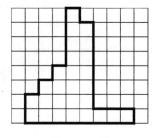

   (A) 27 square units

   (B) 26 square units

   (C) 24 square units

   (D) 29 square units

Practice 31

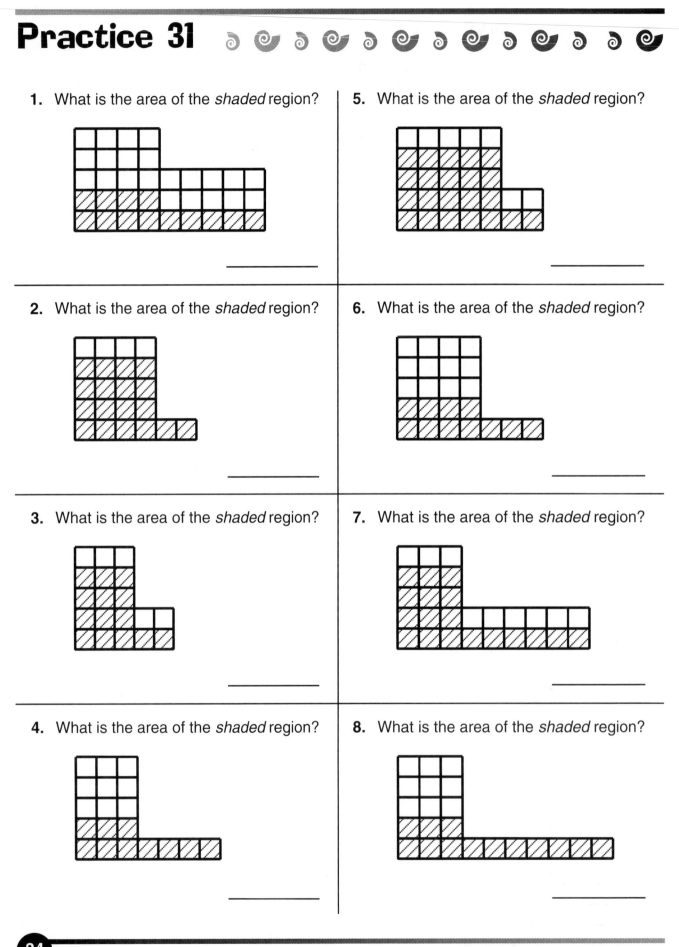

1. What is the area of the *shaded* region?

_____

2. What is the area of the *shaded* region?

_____

3. What is the area of the *shaded* region?

_____

4. What is the area of the *shaded* region?

_____

5. What is the area of the *shaded* region?

_____

6. What is the area of the *shaded* region?

_____

7. What is the area of the *shaded* region?

_____

8. What is the area of the *shaded* region?

_____

Practice 32

1. Which answer gives the perimeter of the figure?

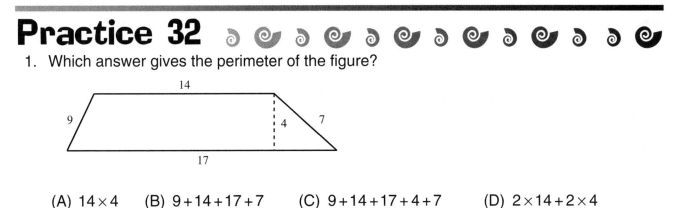

(A) 14×4　(B) $9+14+17+7$　(C) $9+14+17+4+7$　(D) $2 \times 14 + 2 \times 4$

2. Find the perimeter of the rectangle.

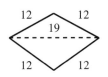

_____

3. Which answer gives the perimeter of the figure?

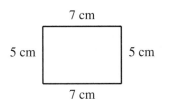

(A) $12+12+12+19+12$　(B) 12×19　(C) $2 \times 12 + 2 \times 19$　(D) $12+12+12+12$

4. What is the area of the figure drawn on the grid?

(A) 23 square units　(B) 22 square units　(C) 25 square units　(D) 20 square units

5. What is the area of the *shaded* region?

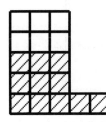

_____

Practice 33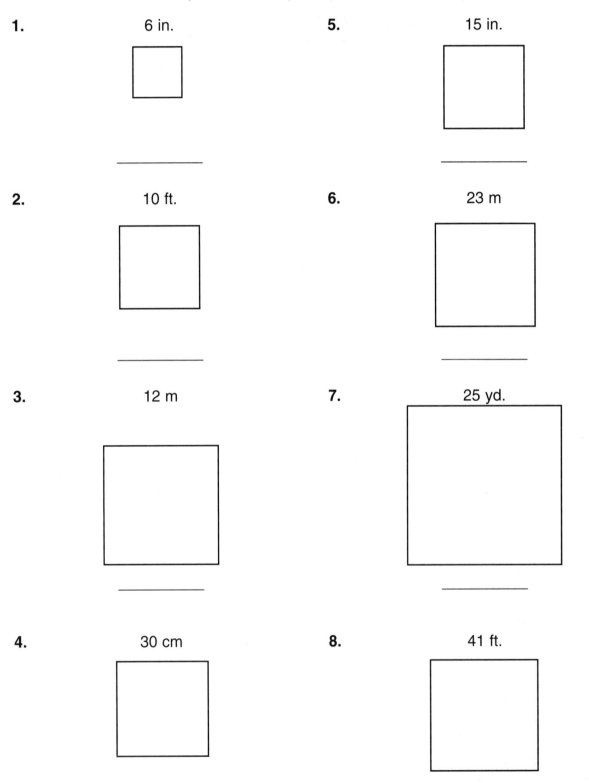

Find the area of each of these figures.

Remember: The area of a square is computed by multiplying the length of one side times itself. **A = s x s** or **A = s²**   (Area = side squared)

1.   6 in.

_____

2.   10 ft.

_____

3.   12 m

_____

4.   30 cm

_____

5.   15 in.

_____

6.   23 m

_____

7.   25 yd.

_____

8.   41 ft.

_____

Practice 34

Find the area of each of these figures.

Remember these formulas.

Area of a rectangle = length times width or base times height

A = *l* x *w* or A = *b* x *h*

Area of a parallelogram = base times height

A = *b* x *h*

1.

5 in.

4 in.

A = _____

2.

9 ft.

5 ft.

A = _____

3.

11 m

7 m

A = _____

4.

15 yd.

6 yd.

A = _____

5.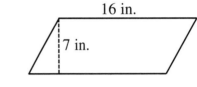

16 in.

7 in.

A = _____

6.

12 cm

11 cm

A = _____

7.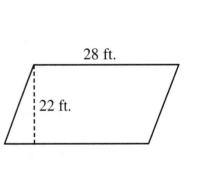

28 ft.

22 ft.

A = _____

8.

50 mm

25 mm

A = _____

Practice 35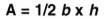

Find the area of each of these figures.

Remember: The area of a triangle is one half the area of a parallelogram or a rectangle. It is 1/2 the base times the height.

$$A = 1/2 \; b \times h$$

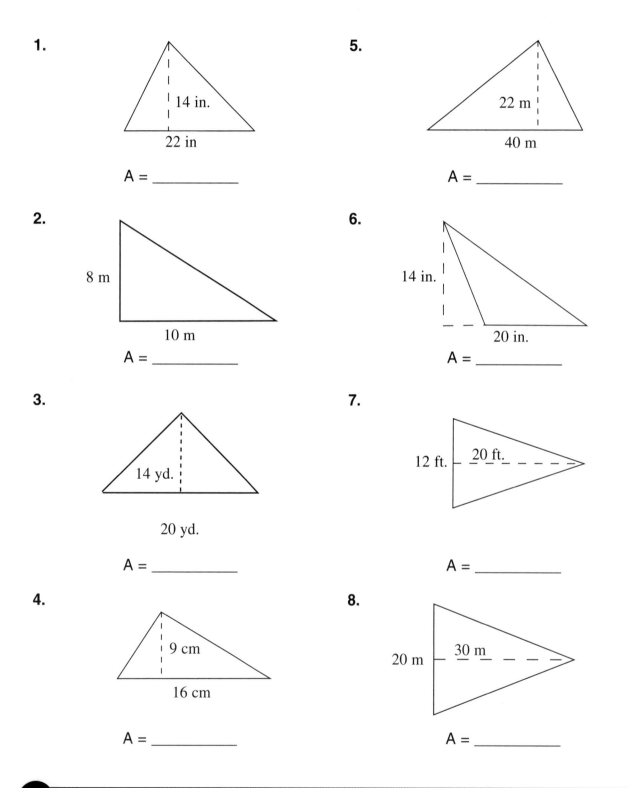

1.

14 in.

22 in

A = _____

2.

8 m

10 m

A = _____

3.

14 yd.

20 yd.

A = _____

4.

9 cm

16 cm

A = _____

5.

22 m

40 m

A = _____

6.

14 in.

20 in.

A = _____

7.

12 ft.   20 ft.

A = _____

8.

20 m   30 m

A = _____

Practice 36

1. The diameter of a circle is 16 centimeters. What is the radius?

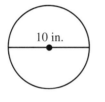

16 cm

_____

2. The diameter of a circle is 10 inches. What is the radius?

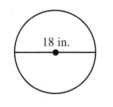

10 in.

_____

3. The diameter of a circle is 18 inches. What is the radius?

18 in.

_____

4. The diameter of a circle is 12 meters. What is the radius?

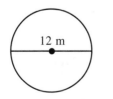

12 m

_____

5. The diameter of a circle is 20 meters. What is the radius?

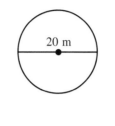

20 m

_____

6. The diameter of a circle is 8 centimeters. What is the radius?

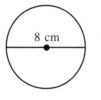

8 cm

_____

7. The radius of a circle is 7 feet. What is the diameter?

7 ft

_____

8. The diameter of a circle is 4 meters. What is the radius?

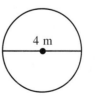

4 m

_____

Test Practice 1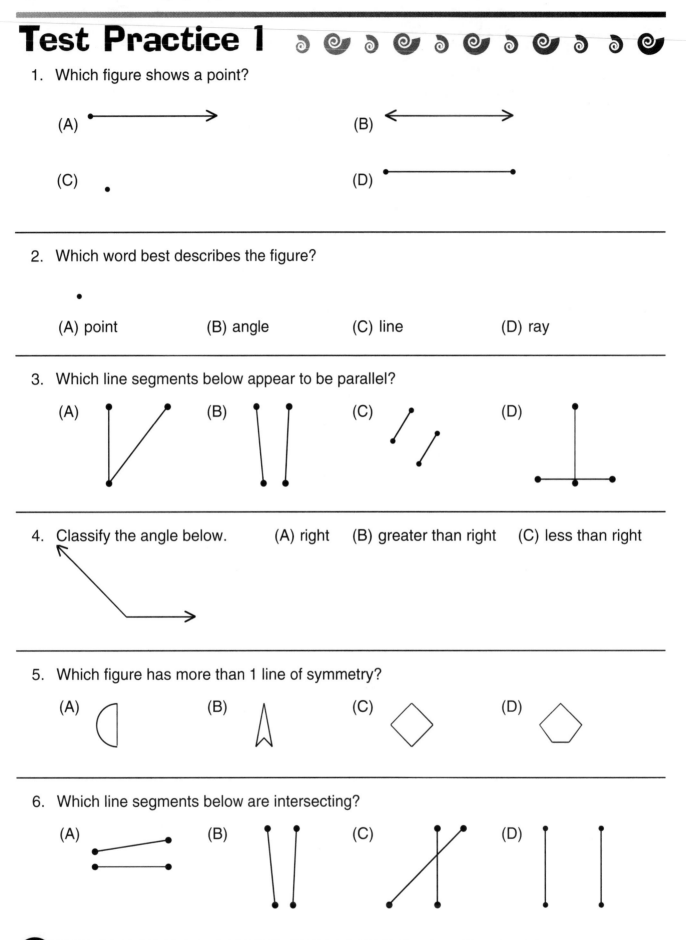

1. Which figure shows a point?

   (A) [ray pointing right]

   (B) [line with arrows both directions]

   (C) .

   (D) [line segment]

2. Which word best describes the figure?

   •

   (A) point   (B) angle   (C) line   (D) ray

3. Which line segments below appear to be parallel?

   (A) [two segments forming V]   (B) [two parallel segments]   (C) [two diagonal segments]   (D) [perpendicular segments]

4. Classify the angle below.   (A) right   (B) greater than right   (C) less than right

   [angle figure]

5. Which figure has more than 1 line of symmetry?

   (A) [half circle]   (B) [arrow shape]   (C) [diamond]   (D) [pentagon]

6. Which line segments below are intersecting?

   (A) [two near-horizontal segments]   (B) [two vertical segments]   (C) [two crossing segments]   (D) [two vertical segments]

Test Practice 2

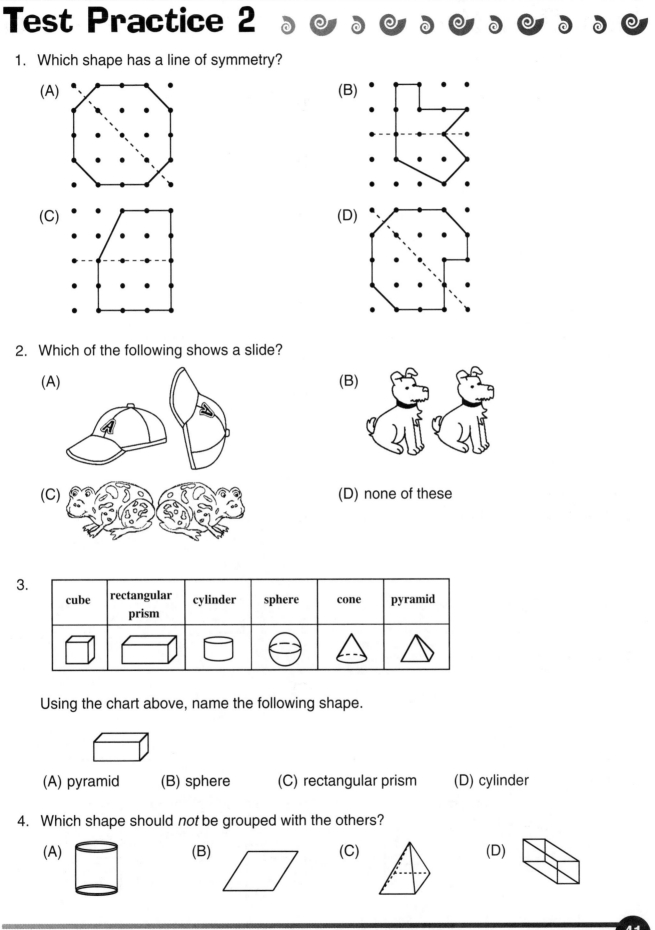

1. Which shape has a line of symmetry?

   (A)

   (B)

   (C)

   (D)

2. Which of the following shows a slide?

   (A)

   (B)

   (C)

   (D) none of these

3.

cube	rectangular prism	cylinder	sphere	cone	pyramid

   Using the chart above, name the following shape.

   (A) pyramid   (B) sphere   (C) rectangular prism   (D) cylinder

4. Which shape should *not* be grouped with the others?

   (A)   (B)   (C)   (D)

Test Practice 3

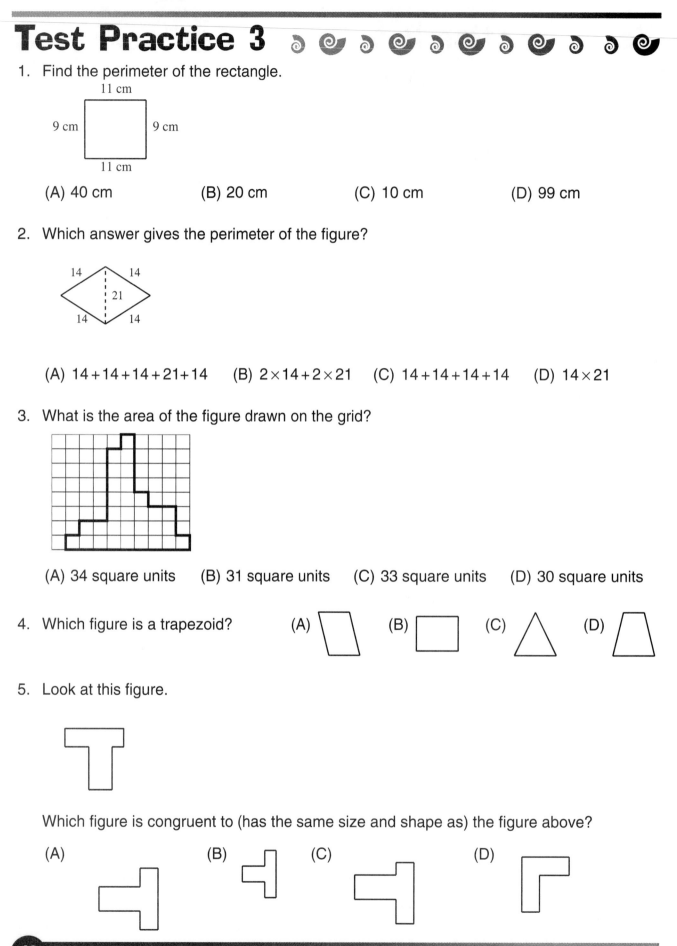

1. Find the perimeter of the rectangle.

   11 cm

   9 cm   9 cm

   11 cm

   (A) 40 cm   (B) 20 cm   (C) 10 cm   (D) 99 cm

2. Which answer gives the perimeter of the figure?

   14   14
   21
   14   14

   (A) $14+14+14+21+14$ (B) $2\times14+2\times21$ (C) $14+14+14+14$ (D) 14×21

3. What is the area of the figure drawn on the grid?

   (A) 34 square units   (B) 31 square units   (C) 33 square units   (D) 30 square units

4. Which figure is a trapezoid?   (A)   (B)   (C)   (D)

5. Look at this figure.

   Which figure is congruent to (has the same size and shape as) the figure above?

   (A)   (B)   (C)   (D)

Test Practice 4

1. If you cut on the dotted line, which one will make two equal parts?

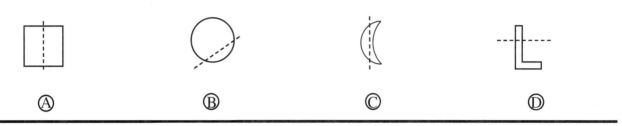

Ⓐ   Ⓑ   Ⓒ   Ⓓ

2. Look at the first shape in the row. Which one of the other shapes looks like it? Fill in the circle under your answer.

Ⓐ   Ⓑ   Ⓒ   Ⓓ

3. If you cut on the dotted line, which one will make two triangles?

Ⓐ   Ⓑ   Ⓒ   Ⓓ

4. How would you find the distance around this figure?

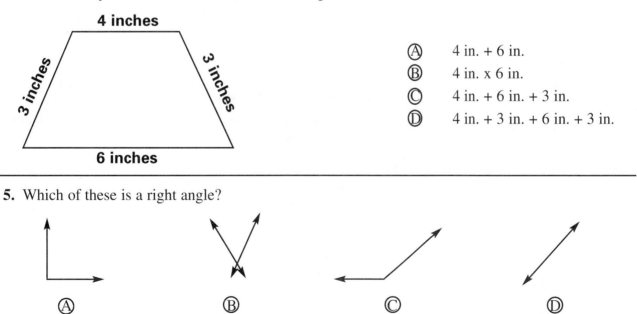

Ⓐ   4 in. + 6 in.
Ⓑ   4 in. x 6 in.
Ⓒ   4 in. + 6 in. + 3 in.
Ⓓ   4 in. + 3 in. + 6 in. + 3 in.

5. Which of these is a right angle?

Ⓐ   Ⓑ   Ⓒ   Ⓓ

Test Practice 5

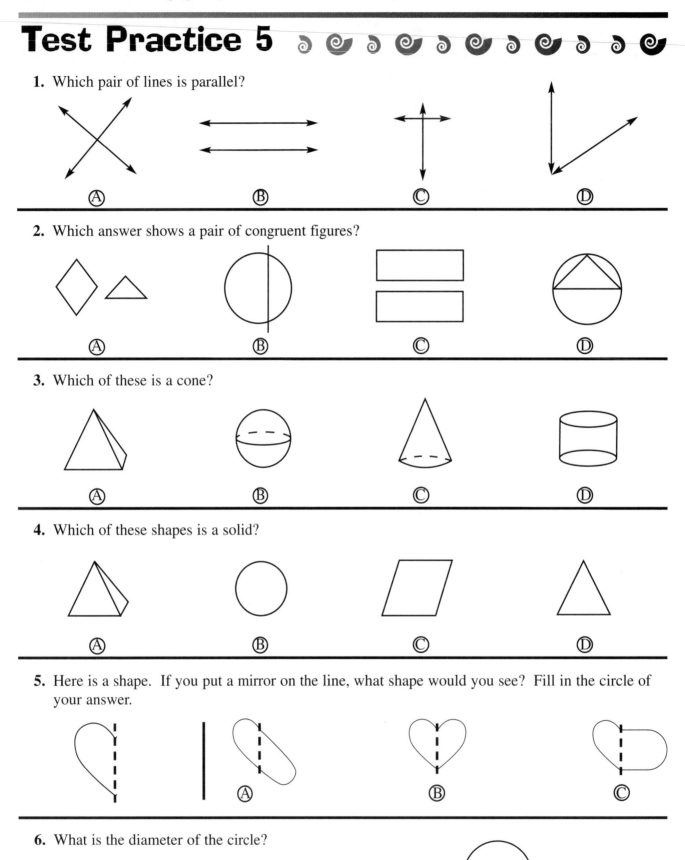

1. Which pair of lines is parallel?

  Ⓐ   Ⓑ   ©   Ⓓ

2. Which answer shows a pair of congruent figures?

  Ⓐ   Ⓑ   ©   Ⓓ

3. Which of these is a cone?

  Ⓐ   Ⓑ   ©   Ⓓ

4. Which of these shapes is a solid?

  Ⓐ   Ⓑ   ©   Ⓓ

5. Here is a shape. If you put a mirror on the line, what shape would you see? Fill in the circle of your answer.

  Ⓐ   Ⓑ   ©

6. What is the diameter of the circle?

  Ⓐ 36 in.   © 12 in.

  Ⓑ 18.35 in.   Ⓓ 24 in.

Test Practice 6

1. Look at the picture of the plate, the clock, and the ball. With which group of shapes do they belong? Fill in the circle of your answer.

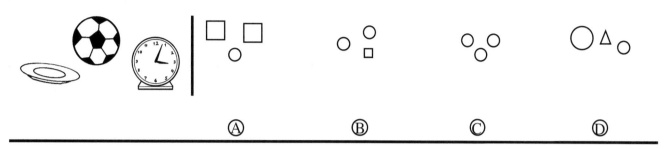

        Ⓐ   Ⓑ   Ⓒ   Ⓓ

2. Look at the picture of the book, the board game, and the mat. With which group of shapes do they belong? Fill in the circle of your answer.

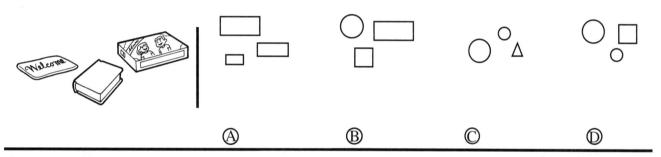

        Ⓐ   Ⓑ   Ⓒ   Ⓓ

3. Here is a group of four small squares that are partly black and partly white. Next is another group of three squares. Choose a square that would make both groups the same.

        Ⓐ   Ⓑ   Ⓒ

4. Look at the ticket. Which word describes the shape of the ticket? Fill in the circle of your answer.

        circle   **square**   **triangle**   **rectangle**

        Ⓐ   Ⓑ   Ⓒ   Ⓓ

Answer Sheet

Test Practice 1	Test Practice 2	Test Practice 3
1. Ⓐ Ⓑ Ⓒ Ⓓ	1. Ⓐ Ⓑ Ⓒ Ⓓ	1. Ⓐ Ⓑ Ⓒ Ⓓ
2. Ⓐ Ⓑ Ⓒ Ⓓ	2. Ⓐ Ⓑ Ⓒ Ⓓ	2. Ⓐ Ⓑ Ⓒ Ⓓ
3. Ⓐ Ⓑ Ⓒ Ⓓ	3. Ⓐ Ⓑ Ⓒ Ⓓ	3. Ⓐ Ⓑ Ⓒ Ⓓ
4. Ⓐ Ⓑ Ⓒ Ⓓ	4. Ⓐ Ⓑ Ⓒ Ⓓ	4. Ⓐ Ⓑ Ⓒ Ⓓ
5. Ⓐ Ⓑ Ⓒ Ⓓ		5. Ⓐ Ⓑ Ⓒ Ⓓ
6. Ⓐ Ⓑ Ⓒ Ⓓ		

Test Practice 4	Test Practice 5	Test Practice 6
1. Ⓐ Ⓑ Ⓒ Ⓓ	1. Ⓐ Ⓑ Ⓒ Ⓓ	1. Ⓐ Ⓑ Ⓒ Ⓓ
2. Ⓐ Ⓑ Ⓒ Ⓓ	2. Ⓐ Ⓑ Ⓒ Ⓓ	2. Ⓐ Ⓑ Ⓒ Ⓓ
3. Ⓐ Ⓑ Ⓒ Ⓓ	3. Ⓐ Ⓑ Ⓒ Ⓓ	3. Ⓐ Ⓑ Ⓒ Ⓓ
4. Ⓐ Ⓑ Ⓒ Ⓓ	4. Ⓐ Ⓑ Ⓒ Ⓓ	4. Ⓐ Ⓑ Ⓒ Ⓓ
5. Ⓐ Ⓑ Ⓒ Ⓓ	5. Ⓐ Ⓑ Ⓒ Ⓓ	
	6. Ⓐ Ⓑ Ⓒ Ⓓ	

Answer Key

Test Practice 1	Test Practice 2	Test Practice 3
1. (A) (B) ● (D)	1. ● (B) (C) (D)	1. ● (B) (C) (D)
2. ● (B) (C) (D)	2. (A) ● (C) (D)	2. (A) (B) ● (D)
3. (A) (B) ● (D)	3. (A) (B) ● (D)	3. (A) ● (C) (D)
4. (A) ● (C) (D)	4. (A) ● (C) (D)	4. (A) (B) (C) ●
5. (A) (B) ● (D)		5. (A) (B) ● (D)
6. (A) (B) ● (D)		

Test Practice 4	Test Practice 5	Test Practice 6
1. ● (B) (C) (D)	1. (A) ● (C) (D)	1. (A) (B) ● (D)
2. (A) (B) ● (D)	2. (A) (B) ● (D)	2. ● (B) (C) (D)
3. (A) (B) (C) ●	3. (A) (B) ● (D)	3. (A) ● (C) (D)
4. (A) (B) (C) ●	4. ● (B) (C) (D)	4. (A) (B) (C) ●
5. ● (B) (C) (D)	5. (A) ● (C) (D)	
	6. (A) (B) ● (D)	

Answer Key

Page 4
1. D
2. C
3. A
4. C
5. A

Page 5
1. B
2. A
3. C
4. A
5. D

Page 6
1. C
2. A
3. D
4. A
5. B
6. D

Page 7
1. D
2. B
3. B
4. D
5. A
6. C

Page 8
1. A
2. B
3. B
4. A
5. C
6. C
7. A
8. A

Page 9
1. A
2. A
3. A
4. C
5. A
6. A

Page 10
1. Yes
2. Yes
3. No
4. No
5. No
6. Yes
7. No
8. No

Page 11
1. B
2. B
3. D
4. B

5. C
6. D
7. B
8. D

Page 12
1. C
2. D
3. A

Page 13
1. D
2. A
3. B

Page 14
1. B
2. B
3. A
4. B
5. B

Page 15
1. slide
2. turn
3. flip
4. slide
5. turn
6. flip

Page 16
1. B
2. C
3. A

Page 17
1. B
2. A
3. D

Page 18
1. acute
2. obtuse
3. acute
4. straight
5. right
6. acute
7. obtuse
8. acute

Page 19
1. right
2. isosceles, acute
3. scalene, obtuse
4. equilateral, acute
5. scalene, obtuse
6. isosceles right

Page 20
1. 45°
2. 60°
3. 40°
4. 45°
5. 130°
6. 120°
7. 45°
8. 120°

Page 21
1. C
2. D
3. B
4. A
5. D
6. D

Page 22
1. A
2. C
3. C
4. A
5. B

Page 23
1. sphere
2. triangular prism
3. cylinder
4. cube
5. cone

Page 24
1. C
2. C
3. D

Page 25
1. C
2. D
3. C

Page 26
1. cube
2. cone
3. sphere

Page 27
1. octahedron
   faces: 8
   edges: 12
   vertices: 6
2. tetrahedron
   faces: 4
   edges: 6
   vertices: 4
3. hexahedron
   faces: 6
   edges: 12
   vertices: 8

4. icosahedron
   faces: 20
   edges: 30
   vertices: 12
5. square pyramid
   faces: 5
   edges: 8
   vertices: 5
6. triangular prism
   faces: 5
   edges: 9
   vertices: 6

Page 28
1. octahedron
2. cylinder
3. hexahedron (cube)
4. cone
5. sphere
6. triangular pyramid
7. rectangular pyramid
8. square pyramid

Page 29
1. D
2. A
3. B
4. C
5. A
6. C

Page 30
1. C
2. C
3. B
4. D

Page 31
1. 37 ft.
2. 40 m
3. 63 in.
4. 63 yd.
5. 23 cm
6. 51 ft.
7. 85 in.
8. 82 m

Page 32
1. 36 in.
2. 39 ft.
3. 80 m
4. 72 yd.
5. 200 ft.
6. 92 in.
7. 175 cm
8. 68 m

Page 33
1. A
2. A
3. A
4. A
5. A
6. A

Page 34
1. 13 square units
2. 18 square units
3. 14 square units
4. 10 square units
5. 22 square units
6. 11 square units
7. 18 square units
8. 13 square units

Page 35
1. B
2. 24 cm
3. D
4. A
5. 11 units

Page 36
1. 36 in.
2. 100 ft.
3. 144 m
4. 900 cm
5. 225 in.
6. 529 m
7. 625 yd.
8. 1,681 ft.

Page 37
1. 20 in.
2. 45 ft.
3. 77 in.
4. 90 yd.
5. 112 in.
6. 132 cm
7. 616 ft.
8. 1,250 mm

Page 38
1. 154 in.
2. 40 m
3. 140 yd.
4. 72 cm
5. 440 m
6. 140 in.
7. 120 ft.
8. 300 m

Page 39
1. 8 cm
2. 5 in.
3. 9 in.
4. 6 m
5. 10 m
6. 4 cm

7. 14 ft.
8. 2 m

Page 40
1. C
2. A
3. C
4. B
5. C
6. C

Page 41
1. A
2. B
3. C
4. B

Page 42
1. A
2. C
3. B
4. D
5. C

Page 43
1. A
2. C
3. D
4. D
5. A

Page 44
1. B
2. C
3. C
4. A
5. B
6. C

Page 45
1. C
2. A
3. B
4. D